Joel Medeiros Bezerra
Caio Sérgio Pereira de Araújo

Groundwater quality in the city of Pau dos Ferros/RN

Joel Medeiros Bezerra
Caio Sérgio Pereira de Araújo

Groundwater quality in the city of Pau dos Ferros/RN

Exploratory study

ScienciaScripts

Imprint

Cover image: www.ingimage.com

This book is a translation from the original published under ISBN 978-3-330-75817-9.

Publisher:
Sciencia Scripts
is a trademark of
Dodo Books Indian Ocean Ltd. and OmniScriptum S.R.L publishing group

120 High Road, East Finchley, London, N2 9ED, United Kingdom
Str. Armeneasca 28/1, office 1, Chisinau MD-2012, Republic of Moldova, Europe
Printed at: see last page
ISBN: 978-620-8-29190-7

SUMMARY

SUMMARY

Groundwater in the municipality of Pau dos Ferros is increasingly exploited to supply the population, as well as local businesses and livestock. The demand for water due to the long drought that has plagued the city is causing concern about the future of water, since there is a lack of management of these resources coming from the population, due to the lack of public policies, which generate uncontrolled withdrawals. As a result, the population uses this water in a disorderly manner without worrying about its quality, since the recharge of these springs has not been happening due to the lack of rainfall, increasing the concentration of minerals and other solids in the catchment wells, which can limit its multiple uses, as well as causing future illnesses to the population. In view of this problem, this work was carried out to assess the quality of groundwater in the municipality of Pau dos Ferros-RN. Groundwater was collected from 18 tube wells in the urban area of the municipality, and physical and chemical parameters were analytically determined in a laboratory at the Federal Rural University of the Semi-Arid Region (UFERSA), in addition to collecting data on the historical use of the wells. Subsequently, the main statistical parameters were analyzed, the linear correlation between the variables studied, box-plot graphs were obtained, as well as the modeling of the surfaces of the variables with thematic isoline maps, in addition to the comparison with the maximum permissible values for human consumption in the federal laws in force in Brazil. The results of the analysis showed that there were wells with water overloaded with calcium and magnesium, thus increasing the electrical conductivity and hardness of the water and consequently making it brackish. As for spatial distribution, well 12 in the Sao Judas Tadeu neighborhood showed the highest levels of calcium and magnesium salts. Finally, the bodies responsible for analyzing groundwater should promote policies for the management and awareness of these resources, in order to prevent future illnesses and even the total scarcity of these resources.

Keywords: Drought, Springs, Water resources, Health risk.

1 INTRODUCTION

Two thirds of the planet's surface is covered by *water,* only a small part of which is made up of fresh water suitable for human consumption, which comes from underground reserves or glaciers. Of all this water available for human consumption, according to the Ministry of the Environment (2007), 96% comes from groundwater, which guarantees the survival of a large part of the world's population, given the limited availability and accessibility of this resource.

In Brazil, around 55% of districts are supplied by groundwater. Cities such as Manaus (AM), Mossoró (RN), Maceió (AL), have all their water needs met through this supply, serving the population and being used in industry, agriculture and leisure (IBGE, 2000).

As the northeastern region of Brazil is characterized by an irregular spatial and temporal distribution of water availability, it is normal for part of the population, those with financial resources, to resort to groundwater abstraction. In Rio Grande do Norte, this is becoming increasingly common, as the state has large aquifers such as the Cristalino, Açu, Jandaira and Barreiras aquifers, which are exploited both legally and illegally.

The quality of groundwater has a high physico-chemical and bacteriological standard, as it is protected, but not totally, from agents of pollution and contamination, and most of the time does not require treatment. While, in terms of quantity, their volume is much greater than that of surface water, their flow is not greatly affected during periods of drought as is the case with surface water, and they have no problems with evaporation losses.

Due to the lack or deficiency in the management of these resources and the high number of users abstracting groundwater, the tendency is for this resource to become scarce, since the amount of water withdrawn is greater than the recharge of its source, as the state is experiencing problems with dry rainfall, which is the main source of recharge to maintain these aquifers.

According to Araùjo (2015), with approximately 60% of the territory covered by crystalline rocks, Potiguar's water table is unable to recover its losses from rainfall, as these surfaces do not absorb water easily. As a result, it is difficult for the State Secretariat for the Environment and Water Resources (SEMARH) to drill wells capable of supplying water of sufficient quality for human consumption. Around 30% of the 320 wells drilled by SEMARH in 2016 failed to spill water.

The state of Rio Grande do Norte has great potential in terms of water resources and underground aquifers, most of which have saline and brackish water in the wells where they are collected, thus requiring a whole apparatus to remove this salt and provide quality water for the population.

In the municipality of Pau dos Ferros, which used to have problems with flooding during rainy periods, the city is now being punished by drought, which is causing the population to despair over the demand for water. As there are no forecasts of rain, the population has begun to exploit groundwater in a disorderly manner by drilling artesian wells in an attempt to meet their needs, disregarding the perception of its sustainability in the face of widespread exploitation in the municipality.

Given the above scenario, the aim of this study was to analyze the quality of groundwater from tube wells in the city of Pau dos Ferros, RN, in densely populated urban areas that suffer from water shortages, in order to meet the water supply needs of the population's multiple uses.

The specific objectives are:

1. Characterize the quality of groundwater from tube wells in the city of Pau dos Ferros;

2. Evaluate the physical and chemical parameters of groundwater;

3. Spatialization of tube wells in the urban area of Pau dos Ferros, in order to relate the spatial distribution of quality to physiographic aspects;

4. Grouping wells based on groundwater quality parameter patterns.

2. LITERATURE REVIEW

2.1 HIDRIC RESOURCES IN RIO GRANDE DO NORTE

According to Silva (2003), water resources are fragile, so they are easily exploited and compromised, whether in terms of quantity or quality, i.e. through other natural or anthropogenic characteristics that change the course of the water, resulting in the reduction of drainage channels and making the lack of this resource even more worrying.

According to Maffei et al. (2009), water is an essential resource, especially in semi-arid regions affected by irregular rainfall, as is the case in the northeast of Brazil, which includes the state of Rio Grande do Norte (RN). Reduced natural availability, combined with poor management of water resources, leads to water contamination, compromising public supply and posing serious risks to people's health and ecosystems.

Water resources are an escape valve that the population has to turn to when it is experiencing a period of severe drought. However, these resources in Rio Grande do Norte are poorly managed, which can lead to a scenario of extreme scarcity, thus jeopardizing the survival of the population and other beings that rely on this resource.

2.2 MONITORING GROUNDWATER RESOURCES

According to Tuinhof et al. (2004), changes in the quantity and quality of groundwater are slow and can only be identified through in-depth monitoring studies over a long period of time. Monitoring provides information for controlling the impacts caused by water extraction and the load of pollutants in the aquifer.

According to UNEP/WHO (1996) and SWRCB (2003), monitoring water quality and quantity are the initial parameters for controlling this resource, helping to make decisions and assessing the veracity of these decisions in the protection, maintenance, improvement and remediation of water resources.

In Brazil, according to the National Water Agency (ANA, 2012), there isn't much information on groundwater, and what does exist is very scattered. As for surface water, the state monitoring networks have around 1,500 monitoring points, in addition to the 1,671 water quality monitoring points that make up the National Hydrometeorological Network.

In Rio Grande do Norte there are several groundwater monitoring points in order to control this resource. The Àgua Azul program consists of the periodic measurement and verification of water quality parameters, used to monitor the current condition, its evolution and control of the quality of the body of water, as well as making it possible to project future situations. It is being carried out in partnership by the Instituto de Desenvolvimento Sustentâvel e Meio Ambiente do Estado do Rio Grande do Norte (IDEMA), Instituto de Gestao de Àguas do Estado do Rio Grande do Norte (IGARN) and Empresa de Pesquisa Agropecuâria do Estado do Rio Grande do Norte (EMPARN), with technical and scientific support from the Universidade Federal do Rio Grande do Norte (UFRN), the State University of Rio Grande do Norte (UERN) and the Federal Institute of Education, Science and Technology (IFRN), which aims to systematically assess the quality of groundwater, providing investigations into potential sources of pollution of water resources during monitoring (IGARN, 2016).

2.3 HYDROGEOLOGICAL FACTORS

Hydrogeology is a branch of hydrology responsible for studying groundwater, especially its relationship with the geological environment, dealing with the geological and hydrological conditions that govern the distribution of groundwater (ABAS, 2016).

According to the Rio Grande do Norte State Secretariat for the Environment and Water Resources (1998), Rio Grande do Norte's underground resources are basically concentrated in five main aquifers, of which the Cristalino, Açu and Jandaira aquifers stand out.

The rocks of the Cristalino formation gave rise to the fissure aquifers, which, due to their geology, do not provide the conditions to supply water, but in most cases, this type of aquifer becomes the only way for small communities to get water, even if it is so little.

According to Costa (2002) the average flow rate of wells in the crystalline zone in the state of Rio Grande do Norte is 2.8 m^3 /h, with 80.1% of the wells having a flow rate equal to or less than 4 m^3 /h, 18.8% being between 4.1 and 16 m^3 /h and only 1% having a flow rate greater than 16 m /h.[3]

The Açu aquifer has an area of 22,000 km^2 and occurs in the Potiguar Basin, where it is confined by the Jandaira limestone. This aquifer is free-form and has a flow of 10 m^3 /h and in its sub-surface the flow can reach up to 200m^3 /h, producing good quality groundwater with a low ionic concentration (SEMARH, 2008; IDEMA, 2008).

The Jandaira aquifer is an important supplier of groundwater with an area of around 15,598 km2. Its thickness varies from 50 to 250 meters horizontally, located in the upper part of the Jandaira Formation. This aquifer is characterized as free-flowing and has flow rates of up to 30 m3/h and an average of up to 3 m3/h. This aquifer has several wells drilled, with an average depth of 8 meters (SEMARH, 2008; IDEMA, 2008).

Meanwhile, the Barreiras aquifer remains confined and semi-confined, and in some areas it is free. With flow rates ranging from 5 to 100m3/h in its wells, its water is of good chemical quality and can be used directly for anthropogenic purposes (SEMARH, 2008; IDEMA, 2008).

The alluvial aquifer is quite dispersed, with sandy sediments deposited in riverbeds. These accumulations have a very ephemeral character, high permeability and considerable recharge, reaching a depth of 7 meters. The quality of its springs is considered good, but it is not widely exploited (SEMARH, 2008; IDEMA, 2008).

2.3.1PAU DOS FERROS BASIN

The Pau dos Ferros Basin, located in the southwestern part of the state of Rio Grande do Norte, covers an area of 65 km^2 and includes four municipalities: Rafael Fernandes, Francisco Dantas, Marcelino Vieira and Pau dos Ferros, of which the municipality of Rafael Fernandes has its administrative headquarters in the basin. Table 1 shows the number of inhabitants per rural and urban area, as well as their territorial area.

Table 1 - Housing and territorial characterization of the municipalities in the Pau dos Ferros basin.

Municipality	Population (inhabitants)			- Area (Km)²
	Urban	Rural	Total	
Rafael Fernandes	1.918	1.898	3.816	5,0
Francisco Dantas	1.493	1.508	3.001	187,9
Marcelino Vieira	3.510	4.557	8.067	324,3
Pau dos Ferros	19.483	2.589	22.072	277,9

Source: CPRM, 2004

2.3.2GEOLOGY AND HYDROGEOLOGY

The Pau dos Ferros Basin is mapped as the Antenor Navarro Formation of Cretaceous age (area of 21 km^2), with colluvial-eluvial cover of Tertiary-Quaternary age occupying an area of 44 km^2 (Figure 1).

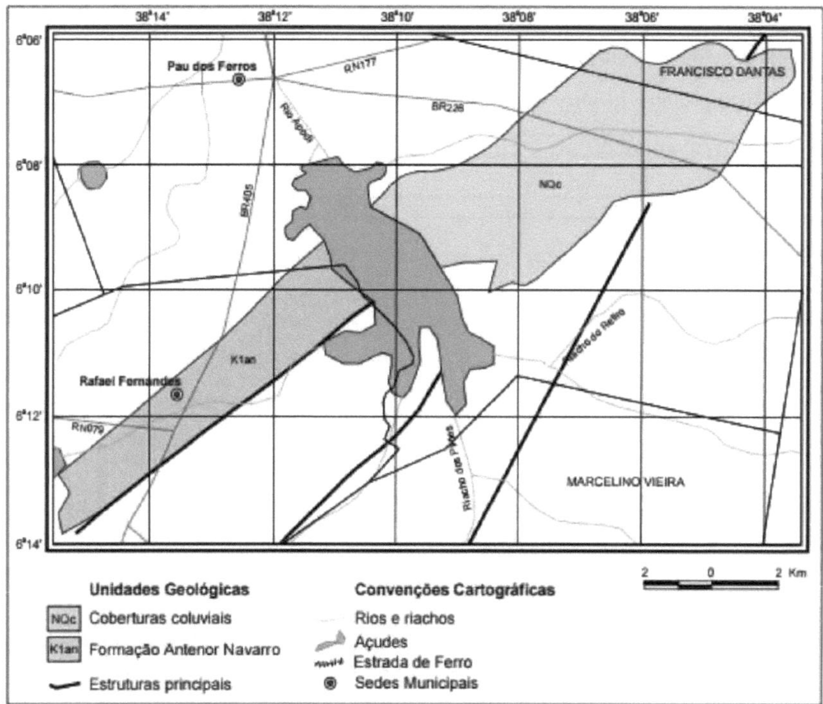

Figura 1 - Geological map of the Pau dos Ferros Basin

Source: CPRM, 2004

The Antenor Navarro Formation is lithologically made up of fine to coarse sandstones, siltstones and claystones (alluvial and fluvial fans intertwined), making it an aquifer with good to medium hydrogeological potential, despite the lack of studies and well data in the area. The colluvial-alluvial cover is made up of sandy, sandy-clayey and conglomeratic sediments, which have low to medium hydrogeological potential due to their small thickness (FEITOSA, 2004).

In terms of quantity, exploitation is incipient and practically non-existent in the northeast, with no information on the few existing wells. However, Figure 2 shows seven privately-owned tube wells with an average depth of 29 meters and an average reported flow rate of 2.65 m /h.[3]

In terms of quality, it can be seen that the area of exposure of the sediment from the roofs makes the aquifer vulnerable, as it facilitates infiltration. There is no further information on the water quality of the existing wells. The seven existing wells have an average electrical conductivity of 1,156 μS⁄cm^2 (CPRM, 2004).

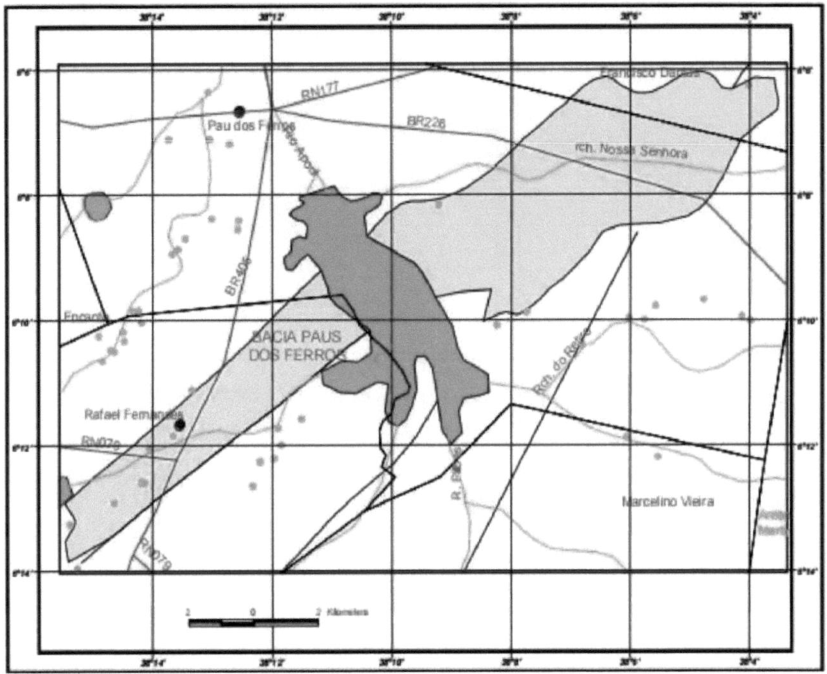

Figura 2 - Map of water points in the Pau dos Ferros Basin

Source: CPRM, 2004

The lithostratigraphic units that occur in the study area (Figure 3) refer to the Caicó Group, which makes up the Gneissic-Migmatitic Complex, formed dominantly by orthognaths, granodiorites, tonalites and granites; The Seridó Group, which includes the Seridó Formation (mica schists in general), the Equador Formation (quartzites) and the Jucurutu Formation (paragneisses); the Umarizal Brasilian Granitoids, made up of granites and various pegmatites; and the Tertiary sedimentary coverings (COSTA et al., 2006).

Infiltration occurs essentially in areas of rock weakness (cracks and fissures), and infiltration rates may be higher under the dominance of alluvial plains and/or soils with vegetation cover.

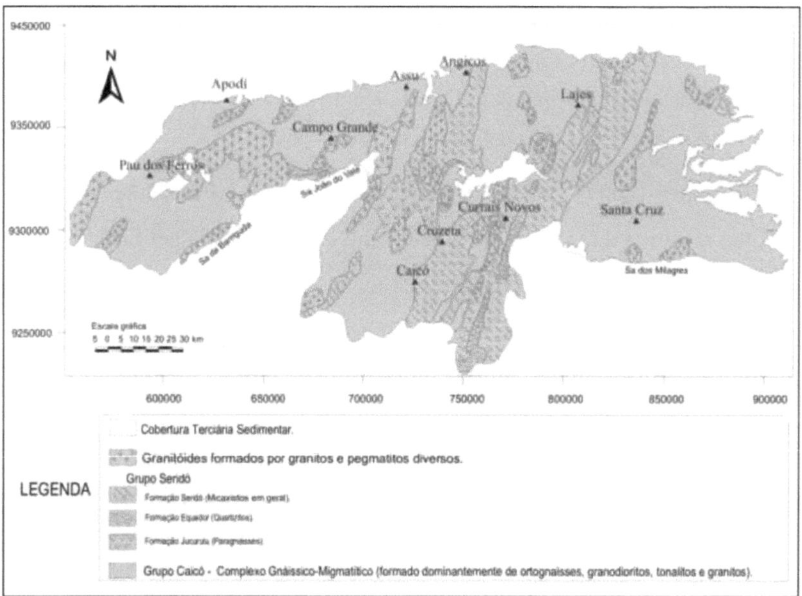

Figura 3 - Geological map of the crystalline in the area studied

Source: DNPM, 1998, simplified by the authors

2.4 GROUNDWATER EXPLOITATION

The exponential growth of the human population has placed enormous demands on water resources, significantly increasing the need for large volumes of water to supply urban populations adequately without causing damage to public health. Urbanization has advanced over water sources and deteriorated surface and underground water supplies (TUNDISI, 2003).

Corroborating this fact, Teixeira et al. (2008) described that the distribution of water throughout the earth is in turn affected by maldistribution, making the search for water resources increasingly precarious, even with a large amount of groundwater existing on earth.

In relation to the increased use of groundwater, the National Water Agency (ANA) states that:

> The use of groundwater has grown rapidly over the last few decades, and there are indications that this trend is set to continue, which explains the continued growth in the number of private companies and public bodies involved in researching and collecting groundwater resources, and the number of people interested in groundwater, both from a technical-scientific and socio-economic point of view, as well as from an administrative and legal point of view (ANA, 2009).

According to Kemper (2007), people who have access to groundwater have no knowledge of the resource that is being exploited and do not want to get rid of it, even if this causes overexploitation. With this lack of knowledge, users need reliable sources and information from agencies to raise their awareness.

2.5 GROUNDWATER QUALITY

The amount of good quality fresh surface water is decreasing every day due to anthropogenic action, combined with climate issues. As a result, society is turning to other sources of water, such as groundwater. As these waters go through their natural filtering process, they provide good quality for use. This does not apply to all groundwater and does not require specific treatment.

Groundwater is a resource that people use very frequently. Groundwater can be collected from the confined or artesian aquifer, which is located between two almost impermeable layers, making it difficult to contaminate, or from the unconfined or free aquifer, which is located on the surface and is therefore more subject to contamination from various polluting sources. Due to its low cost and ease of drilling, water abstraction from the free aquifer, although more vulnerable to contamination, is more frequently used in Brazil (FOSTER, 1993; SILVA, 1999).

In general, groundwater does not need initial treatment, as it is given potable status when it passes through filtering processes in the rocks and subsoil along its course, promoting its cleaning and purification, making it a good quality, low cost water that can be exploited in both rural and urban areas (OLIVEIRA; LOUREIRO, 1998).

Still on the subject of the exploitation of groundwater, it can be seen that its frequent use, whether to meet the needs of the human population or for other purposes, is due to its quality compared to surface water, which is due to the greater natural filtration of the soil that protects aquifers from contamination, which is not the case with surface water (LARINI, 2013).

Several factors can compromise the quality of groundwater. The final disposal of domestic and industrial sewage in cesspits and septic tanks, and the inadequate disposal of solid urban and industrial waste, are all sources of contamination of groundwater by pathogenic bacteria and viruses, parasites, organic and inorganic substances (SILVA, 2003; ARAUJO, 2015).

Agriculture is the activity that pollutes groundwater the most. According to Hirata (2003), these activities pollute the water table, which in turn contaminates groundwater with inorganic substances such as insecticides and fertilizers, making it unsuitable for direct use.

In Brazil, the quality standard for water for human consumption, approved in Ordinance No. 2.914, December 12, 2011, of the Ministry of Health, defines the maximum permissible values (VMP) for the bacteriological, organoleptic, physical and chemical characteristics of drinking water (BRASIL, 2000). According to art. 4 of this ordinance, water of good quality for human use and consumption is considered to be that which meets the drinking water standard and whose microbiological, physical and chemical parameters are at their optimum values and do not pose any risk to human health.

3. METHODOLOGY

3.1.1STUDY AREA

The municipality of Pau dos Ferros is located in the West Potiguar mesoregion, bordering the municipalities of Francisco Dantas, Sao Francisco do Oeste, Marcelino Vieira, Rafael Fernandes, Antônio Martins, Serrinha dos Pintos and Encanto and the state of Cearà, covering an area of 277 km^2 , on the Pau dos Ferros sheet (SB.24-Z-A-II), on a scale of 1:100,000, published by SUDENE (BELTRÂO, 2005). Figure 4 shows the road access map for the municipality of Pau dos Ferros.

The municipality is characterized by a hot, semi-arid climate with rainy seasons tending towards autumn. From January 2016 to November 2016, the accumulated rainfall was 415mm, compared to the entire year of 2015 when the rainfall was 469mm, which is below the dry level of 490.34mm (EMPARN, 2016).

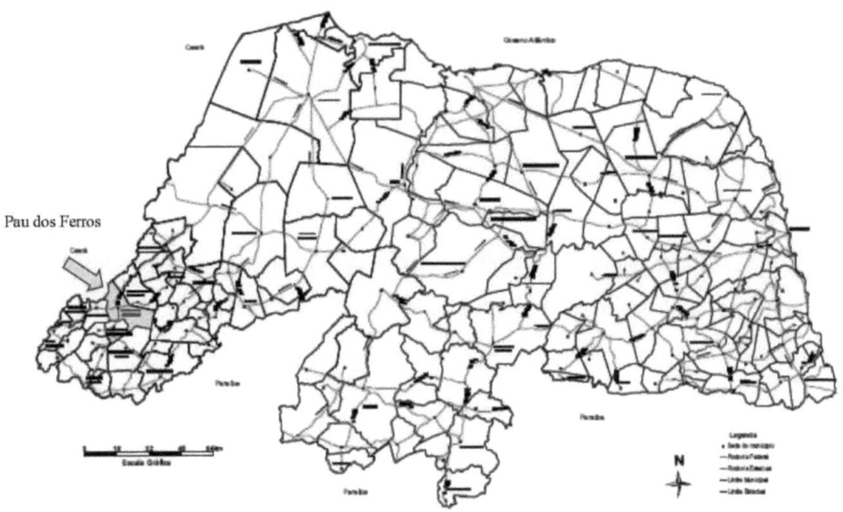

Figure 4 - Road Access Map

Source: CPRM, 2004.

Figure 5 shows the maximum and minimum rainfall records for the months of March and November 2009 in Rio Grande do Norte. The highest temperatures in Rio Grande do Norte can be found in the Alto Oeste Potiguar region, in the semi-arid northeast.

The municipality of Pau dos Ferros is geographically located between mountains, which causes the average air temperature to rise much higher in summer. According to EMPARN (2009), Pau dos Ferros has an average annual air temperature of around 28.1°C, with a maximum of 36°C and a minimum of 21°C (Figure 6).

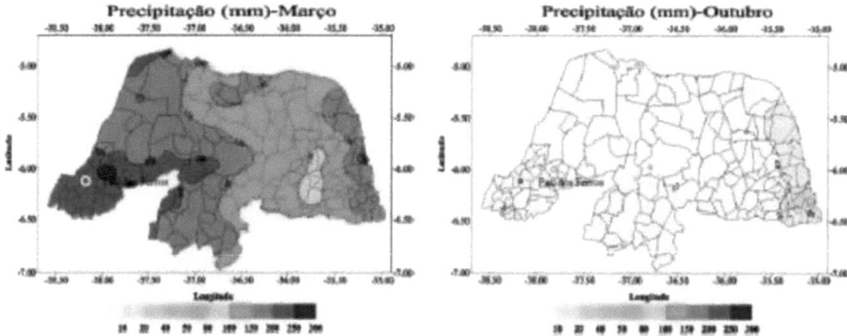

Figure 5 - Regionalization of maximum and minimum rainfall for the months of March and November 2009, for Rio Grande do Norte.

Source: EMPARN, 2009

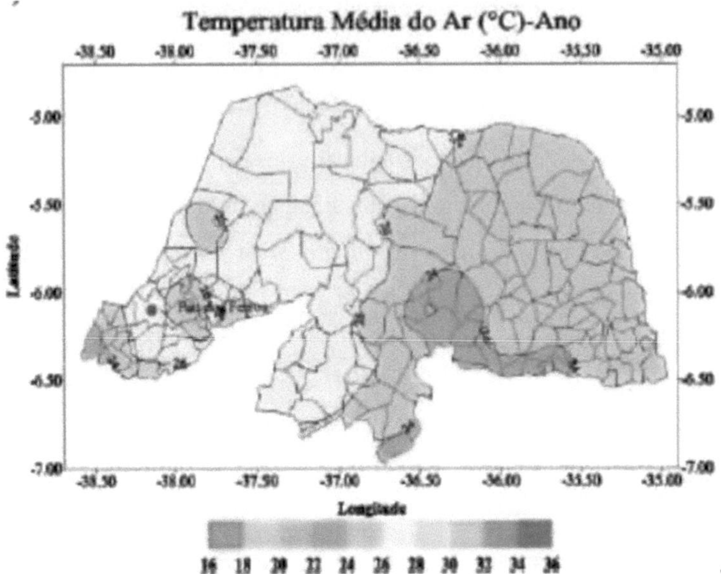

Figura 6 - Regionalization of the average air temperature for 2009 in Rio Grande do Norte

Source: EMPARN, 2009

The predominant biome in Pau dos Ferros is the Caatinga and is associated with the Caatinga Hiperxerofita sub-group. According to Fernandes (2007), *there is* an intense influx of light into the arboreal and deciduous vegetation, creating a bright and sunny climate compared to the dark and closed conditions of the forests.

In the municipality of Pau dos Ferros, the predominant formation is hyperxerophytic with a dry character and an abundance of cacti and small plants. Among the various species found there, the ones that stand out the most are jurema-preta, mufumbo, xique-xique, marmeleiro and facheiro (IDEMA, 2009).

The soils of Pau dos Ferros basically consist of three types, Argissolo, Chernossolo and Luvissolo, of which the most predominant in the region is the equivalent eutrophic red-yellow Podzolic (Argissolos), with the main characteristics of high fertility, medium and medium gravelly texture, markedly drained and gently undulating relief, see Figure 7 (BELTRÂO, 2005).

According to the same author, agricultural use without adequate irrigation to meet the soil's needs is restricted to drought-resistant crops, and erosion control is recommended. Its agricultural suitability is very limited to the cultivation of crops, but it is effective for crops with long cycles such as cotton, sisal, cashew and coconut, with a small area set aside for the preservation of fauna and flora or for some kind of recreation (BELTRÂO, 2005).

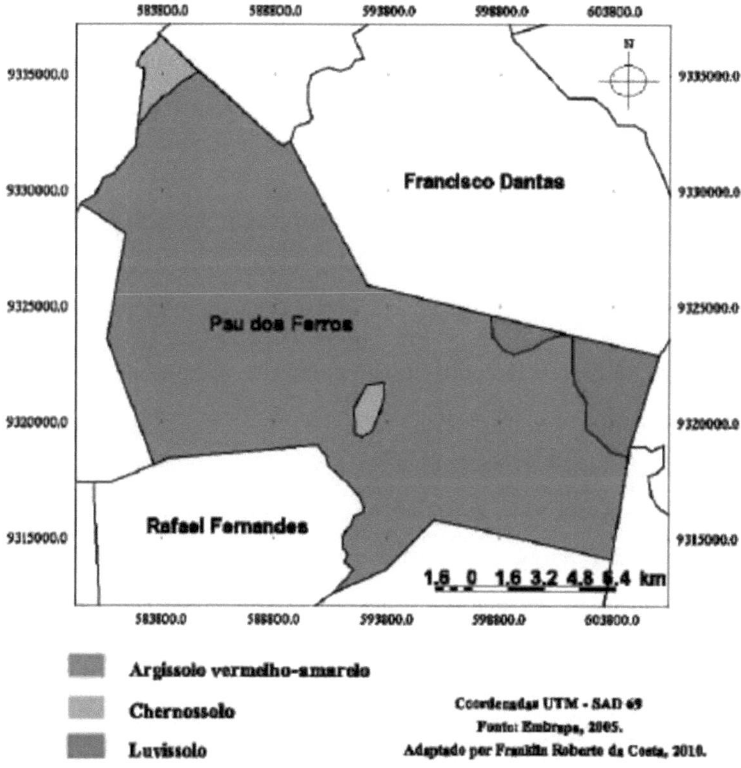

Figura 7 - Soil Chart of the Municipality of Pau dos Ferros Source: Adapted from EMBRAPA (2005).

According to the 2015 census of the Brazilian Institute of Geography and Statistics (IBGE), the total resident population in the municipality of Pau dos Ferros is 27,775 inhabitants, of which 13,516 are male, or 48.71%, and 14,229 are female, or 51.28%. The literate resident population is 21,011, or 75.73% of the population. Its estimated demographic density in 2010 was 106.73 (inhabitants/km^2) and its HDI in 2010 was 0.678 (IBGE, 2010).

Statistics from the central register of companies in 2014 show that the municipality has 685 companies that operate directly in the economy, with 703 units within the city, employing 3,435 people on a fixed salary and a total of 4,212 people who are in the labor market with or without a formal contract, earning an average of two minimum wages (IBGE, 2014).

According to Beltrao (2005), the municipality carries out various activities that make the town's economy spin, among them agriculture, extractivism and local commerce.

3.2 FIELD PROCEDURES

The fieldwork was carried out in the consolidated urban area of the municipality of Pau dos Ferros-RN, from September to October 2016. The tubular wells were located using a coordinate system with the aid of a Garmin 78H Global Positioning System (GPS).

In households where the water was collected by pump, the samples were taken at a point where the water came from the well before reaching any reservoir. In households where the water was collected by bucket pulled by rope, the samples for analysis of physico-chemical parameters were collected in a virgin container (bucket that had never been used), pulled by a rope that was also virgin (never used) and transported to the appropriate containers for each analysis. All collection containers were labeled with the sample data. Collection forms were filled in with the data relating to the sample collected (address, time, sample number, weather conditions and type of collection - pump or bucket), which were sent to the laboratories with the packaged samples, for subsequent collection of the variables at the Engineering Applied Chemistry Laboratory of the Federal Rural University of the Semi-Arid (UFERSA), in accordance with the technical guidelines of the Ministry of Health, in its Ordinance 2.914/2011, following the methods used in the Standard Methods for the Examination of Water and Westwater, 20th edition.

As not all the existing artesian wells have a water abstraction license or are even registered with the Pau dos Ferros city council, the sample area was made up of wells that could be identified through *on-site* visits and accessibility provided by the owners. Figure 8 shows the sites sampled.

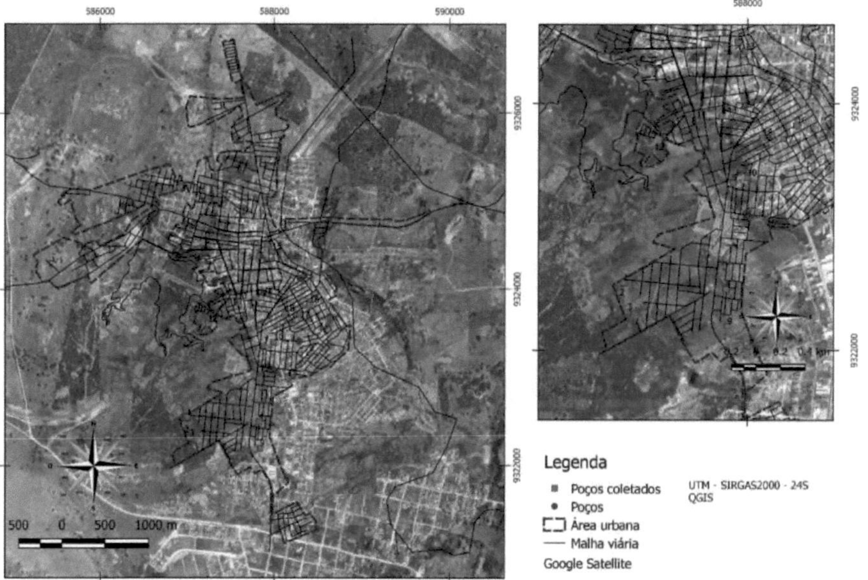

Figure 8 - Location of the wells used for groundwater sampling in the urban area of Pau dos Ferros.

Source: Prepared by the authors

Data was collected on both physical variables (depth, flow, time of use, altitude) and chemical variables (turbidity, conductivity, pH, total solids and total dissolved solids, hardness, calcium, magnesium), in order to gain a better understanding of the assessment and the purpose of its use.

The values for depth, flow rate and time of use were obtained from information provided by the well managers for the private wells, while for the public wells the information was provided by the municipality's Department of Agriculture. All the samples were collected by a single person, qualified and experienced in the procedures for collecting water for analysis. The altitude was obtained with the help of a Garmin 78H GPS, which provided UTM georeferenced coordinates in addition to the altitude.

The pH values were obtained using a pHmeter, brand pHmeter, JK- PHM-005. Turbidity was obtained using an Adamo turbidimeter. Electrical conductivity was measured using a Conducter meter CD-860.

Total solids were determined using the gravimetry method according to Standard Methods (APHA, 1998). Total dissolved solids were obtained using the empirical relationship proposed by APHA et al. (1992) between conductivity (C, µmho/cm) and the concentration of total dissolved solids (STD, mg/l).

The titrimetric determination method was used to obtain the hardness and calcium concentrations, and the concentrations were obtained using Equation 1, where N_{EDTA} is the molarity of the EDTA reagent, and $VEDTAgasto$ is the volume of EDTA used in the titration. The formulas for hardness and calcium were taken from the Manual of Laboratory Procedures and Techniques for the Analysis of Sanitary and Industrial Water and Sewage (2004).

$$mgCaCO_3/L = \frac{N_{EDTA} * V_{EDTAgasto} * 50000}{Volume\ da\ amostra} \qquad \text{(Equation 1)}$$

Magnesium was obtained from Equation 2, using the difference between the volume of EDTA used to titrate hardness and calcium.

$$mgCaCO_3/L = \frac{EDTA\ gastos\ na\ dureza - EDTA\ gastos\ no\ cálcio * 0{,}05 * 50000}{Volume\ da\ amostra}. \qquad \text{(Equation 2)}$$

3.3 OFFICE PROCEDURES

The office procedures were carried out at the Federal Rural University of the Semi-Arid, Multidisciplinary Center of Pau dos Ferros, and the procedures for analytical determination of the parameters to be evaluated were carried out in the analytical chemistry laboratory, where, after obtaining the parameters mentioned, all the data tabulation was carried out, along with descriptive statistical analysis, production of box-plot graphs, obtaining linear correlation of the physical and chemical variables, using Microsoft Office Excel, as well as spatializing the georeferenced data in thematic isoline maps, through surface modeling processes, using the QGIS 2 program.14.7.

Next, the association between the characteristics of the wells (depth, type of catchment, nearest well-trench distance and area studied) and the water quality parameters surveyed was evaluated.

4 RESULTS AND DISCUSSION

The Pau dos Ferros Basin's hydrogeological potential is not as great as that of the large sedimentary basins in the Northeast, but it does represent a palliative form of water supply for local, mainly rural, populations.

The wells studied were shallow, mechanically drilled, located in the free aquifer, above the relatively impermeable rock layer that protects the aquifer from infiltration and contamination.

Table 2 shows the results obtained through analytical determinations in the laboratory, which show the main descriptive statistics of the groundwater quality data evaluated.

Evaluating the values in Table 2 for mean and variance, it can be seen that there is stationarity for the parameters of pH, electrical conductivity and time of use, since the values are very close, while there was a discrepancy for the values of Hardness, Calcium and Magnesium, due to the high values obtained in the analyses.

In terms of variance, attention should be paid to the values for turbidity, pH, conductivity and time in years, which remained close, obtaining a low variance with little variability in the data. For the hardness, calcium and magnesium data, there was a large dispersion of data, with high variations, due to the high values in the concentrations obtained in the analyses, with a focus on the concentrations of calcium and magnesium ions, found in excess in most wells.

The amplitude found in Table 2 was close for pH, conductivity and time in years, identifying little distance between the data, with a discrepant value for the turbidity variable, due to the quantities of ions found in well 10, being reinforced by the value of asymmetry and kurtosis, which can be characterized as a non-normal distribution (lognormal). The amplitude for Calcium Magnesium and Hardness behaved highly due to the distance from one value to another in the variables.

Meanwhile, the values for ST and STD were very low, which is related to the amount of salts present in the samples, thus demonstrating the homogeneity in the behavior of the total dissolved solids of the waters of the crystalline rock region as a whole. This suggests similar conditions of recharge, storage and underground flow dynamics in the fissure environment.

The values of the coefficient of variation make it possible to accurately interpret the accuracy of the values obtained. According to Gomes (1984), values of less than 10% are described as low, giving values of 10-20% medium, 20-30% high, and over 30% very high accuracy.

Table 2 shows that the lowest CV value was found for pH (4.36%), while the highest CV was obtained for turbidity (354.90%). Similar values were found in the Research and Development Bulletin (Regionalization of groundwater quality parameters for irrigation in the state of Sergipe) developed by EMBRAPA (2008).

Hardness expressed in mg/L had a minimum value of 142.50 mg/L, which is classified as moderately soft water, while its maximum value was 987.50 mg/L, classifying it as very hard water, as its value was greater than 200 mg/L, according to the values provided by Custódio & Llamas (1983).

Table 2 - Descriptive statistics for groundwater quality parameters

Parameters	TURB (NTU)	pH	COND (mS/cm)	DUR (mg/L)	Ca (mg/L)	Mg g/(mL)	Q (L/h)	PROF (m)	TU (years)	ALT (m)	STD (mg/L)	ST (mg/L)
NA	18	18	18	18	18	18	18	18	18	18	18	18
Minimum	0,00	6,51	1,00	142,50	75,00	67,50	500,00	42,00	0,60	167,00	6,40E-04	1,71E-02
Maximum	5,80	7,64	5,38	987,50	450,00	537,50	3500,00	100,00	3,00	307,00	3,44E-03	9,60E-02
Range	5,80	1,13	4,38	845,00	375,00	470,00	3000,00	58,00	2,40	140,00	2,8E-03	7,89E-02
Average	0,38	6,95	2,20	343,33	168,19	175,14	1520,17	58,06	1,93	212,11	1,41E-03	3,62E-02
Median	0,00	6,90	1,60	242,50	120,00	126,25	1200,00	55,00	2,05	211,50	1,02E-03	2,41E-02
Variance	1,85	0,09	1,76	60275,74	12605,74	18734,91	849274,62	210,53	0,74	930,22	7,2E-07	6,07E-04
Standard Deviation	1,36	0,30	1,33	245,51	112,28	136,88	921,56	14,51	0,86	30,50	8,49E-04	2,46E-02
CV (%)	354,90	4,36	60,27	71,51	66,75	78,15	60,62	24,99	44,70	14,38	60,27	68,05
Asymmetry	4,16	0,61	1,61	1,70	1,52	1,75	0,88	1,58	-0,11	1,73	1,61	1,66
Kurtosis	17,46	0,06	1,40	1,99	1,36	2,07	-0,37	3,10	-1,34	4,98	1,40	1,50
D	0.416Ln	0,13n	0,341n	0,316n	0,274n	0,336n	0,247n	0,225n	0,17n	0,213n	0,341n	0,359n

TURB - Turbidity; pH - Hydrogenionic Potential; COND - Electrical Conductivity; DUR - Total Hardness; Ca - Calcium; Mg - Magnesium; Q - Flow Rate; PROF - Depth; TU - Time of Use; ALT - Altitude; STD - Total Dissolved Solids; and ST - Total Solids.

NA - Number of samples; Min - Minimum; Max - Maximum; CV - Coefficient of variation (%); D - Maximum deviation from the normal distribution using the Kolmogorov-Smirnov test with a probability of error of 0.05; n - Normal Distribution; Ln - Lognormal Distribution

Source: Prepared by the authors

Figure 9 shows the box-plots of the variables under study, divided into sets based on their respective orders of magnitude. An analysis of Figure 9a shows that the medians of the total dissolved solids and total solids parameters were equal to zero, due to the fact that the amount of solids contained in the samples was small.

For turbidity, there was a discrepant value (*outliers*), which is due to the lack of recharge of the water table, resulting in an increase in salt particles or suspended solids present in the water for the *outlier*. Turbidity, suspended material in the water, can attach itself to existing pathogens, protecting them and even hindering the action of chlorine on them.

When analyzing the ST results, turbidity remained low because it contained few suspended solids. The flow rate showed a dispersion of data due to the geological formations of the city and the fact that the municipality is in a period of drought, with no recharge, increasing the variability between the flow rates.

For the analysis in Figure 9b, pH and time of use showed much higher medians than electrical conductivity. According to the Ministry of Health's Ordinance 2914/2011, the pH of water is recommended to be between 6.0 and 9.5, so all the wells are within the standards in force, although there is a tendency towards alkaline water as they are slightly higher than the reference value of 7.0.

The alkalinity of water can contribute to the formation of crusts on the structures of hydraulic installations, adding constituents to the water. The biggest changes in this indicator are caused by domestic and industrial waste (DERISIO, 1992). The final disposal of domestic and industrial waste in cesspits or septic tanks may be contributing to the alkalinity of groundwater in the areas investigated, with the percolation of various substances.

The COND values indicate the presence of a high concentration of salts due to the high amount of Ca^{2} + and Mg^{2} + ions.

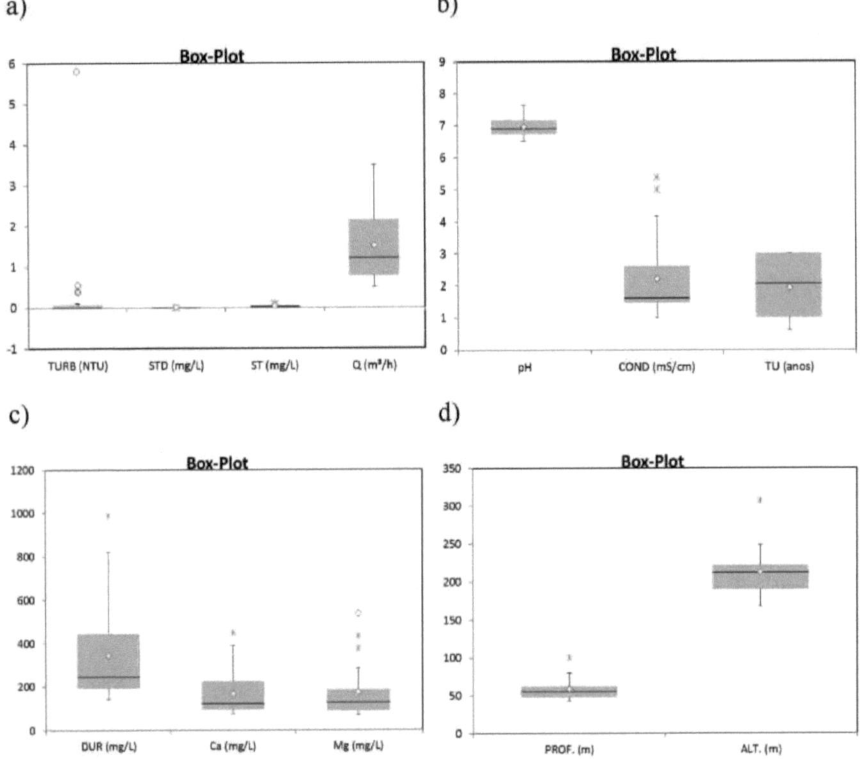

Figure 9 - Box-Plot graphs with a set of variables: a) Turbidity, Total Dissolved Solids, Total Solids and Flow Rate; b) pH, Electrical Conductivity and Time in Years; c) Hardness, Calcium and Magnesium; and d) Depth and Altitude Source: Prepared by the authors

The three variables in Figure 9c have total data affinity in relation to the concentration of salts in the water. Mg showed a discrepant value (*outliers*), which exceeded the maximum value. Hardness and Ca, on the other hand, showed a much greater dispersion of data than Mg. The dispersion of hardness may be due to the amount of Ca and Mg ions dissolved in the sample, which are ions in its composition. Ca and hardness had one extreme value, while Mg had two extreme values. These values could be the result of high spatial variability in groundwater recharge, as well as geological formation.

Figure 9d shows the box-plots for the variables depth and altitude, from which it can be seen that there were no discrepant values in the results. The dispersion of the altitude box-plot was much greater than that of depth, which remained proportional, with both variables showing extreme values. Depth was the result of the need to drill deeper, due to the difficulty of exploiting water of good quality and significant quantity, because of the scarcity and lack of recharge from rainfall, while altitude was due to the geological and topographical formation of the municipality.

Table 3 shows the linear correlation matrix between the variables under study and is evaluated according to the classification proposed by Santos (2007). The correlation values presented coincide with the dispersion graphs of the behavior of these elements (Figure 9).

A similar behavior was observed between the distribution of the concentrations of the variables, especially with values of $|r| > 0.8$ where they showed strong positive correlations. Correlations of $0.5 < |r| < 0.8$ were classified as moderate; $0.1 < |r| < 0.5$ weak; $0 < |r| < 0.1$ very weak, according to Santos (2007).

Table 3 - Linear correlation between the parameters evaluated

	TURB (NTU)	PH	COND (mS/cm)	DUR (mg/L)	Ca (mg/L)	Mg (mg/L)	Q (L/h)	PROF (m)	TU (years)	ALT (m)	STD (mg/L)	ST (mg/L)
TURB (NTU)	1,000											
pH	-0,395	1,000										
COND (mS/cm)	-0,171	-0,399	1,000									
DUR (mg/L)	-0,115	-0,487	**0,979**	1,000								
Ca (mg/L)	-0,119	-0,485	**0,951**	**0,982**	1,000							
Mg (mg/L)	-0,108	-0,476	**0,976**	**0,988**	**0,941**	1,000						
Q (L/h)	0,344	0,062	-0,278	-0,290	-0,288	-0,284	1,000					
PROF. (m)	0,020	-0,320	0,471	0,444	0,361	**0,500**	-0,336	1,000				
TU (years)	0,323	0,254	-0,452	-0,461	-0,447	-0,460	0,176	-0,459	1,000			
HEIGHT (m)	0,100	-0,360	-0,320	-0,201	-0,170	-0,221	-0,130	0,032	0,124	1,000		
STD (mg/L)	-0,171	-0,399	**1,000**	**0,979**	**0,951**	**0,976**	-0,278	**0,471**	-0,452	-0,320	1,000	
ST (mg/L)	-0,135	-0,432	**0,996**	**0,984**	**0,962**	**0,976**	-0,259	**0,446**	-0,450	-0,292	**0,996**	1,000

TURB - Turbidity; pH - Hydrogenionic Potential; COND - Electrical Conductivity; DUR - Total Hardness; Ca - Calcium; Mg - Magnesium; Q - Flow Rate; PROF - Depth; TU - Time of Use; ALT - Altitude; STD - Total Dissolved Solids; and ST - Total Solids.

In the hardness concentration *there are* calcium salts present which prevent the soap from dissolving in the water. While the correlation between Mg and electrical conductivity is due to the high concentrations of Mg dissolved in the water in which it is detected in the electrical conductivity test. The correlation between Mg and hardness is also due to the high concentrations of magnesium in the water, in which hardness detects this presence, which can lead to water that is overly concentrated in salts, compromising its quality and limiting its uses, such as human and animal consumption, as well as the dissolution of soap, causing pipes to clog.

The relationship between STD and electrical conductivity indicates a strong integration of the data, due to the amount of solids such as Ca, Mg and others present in the water and being perceived in the electrical conductivity reading. The STD shows all the material dissolved in the water, whether of ionic or colloidal origin.

Figures 10 and 11 show the thematic isoline maps for the parameters assessed. Meanwhile, Figure 12 shows the thematic map of the multiple uses of the water collected from the tube wells. It can be seen that the water is used for various purposes, including frequent use for commercial and domestic activities.

Figure 10a shows that the highest point is located in the city center, at 307 m. This value is due to the characteristics of the municipality, which is located in a region of mountainous terrain.

Figures 10b, 10c and 10d show the spatial distribution of calcium ion concentrations, electrical conductivity and hardness, where the highest concentration was found in well 12, located in the Sao Judas Tadeu neighborhood.

The high calcium value of around 449mg/L is associated with the dry season in the region, so the lack of recharge of the springs due to the drought and the lack of rain in the region intensifies the dissolution of calcium salts present in the rocks which is associated with the high salinity of the water, also increasing the electrical conductivity which obtained a value of 5.37mS/cm, a variable which measures the content of salts present in the water.

Therefore, the high amount of calcium ions will also increase the hardness of the water, which had a value of 987mg/L, exceeding the value established by the Ministry of Health's ordinance 2914/2011, which is 500mg/L.

Figures 10e and 10f show the turbidity and flow rate maps, which registered their extreme points at well 10, with a turbidity value of 5.8 NTU, while the flow rate was 3,500L/h at well 5, located in the city center. Turbidity is due to the presence of solid materials suspended in the water, and the value obtained was due to the amount of salts in suspension in well 10, which is outside the drinking water standards in force according to Ministry of Health ordinance 2914/2011, which limits a maximum value of 5.0 NTU and can leave the water with poor organoleptic properties, such as bad taste and bad smell.

Well 5 in the city center refers to a public supply point, which meets the needs of various users who may require this resource.

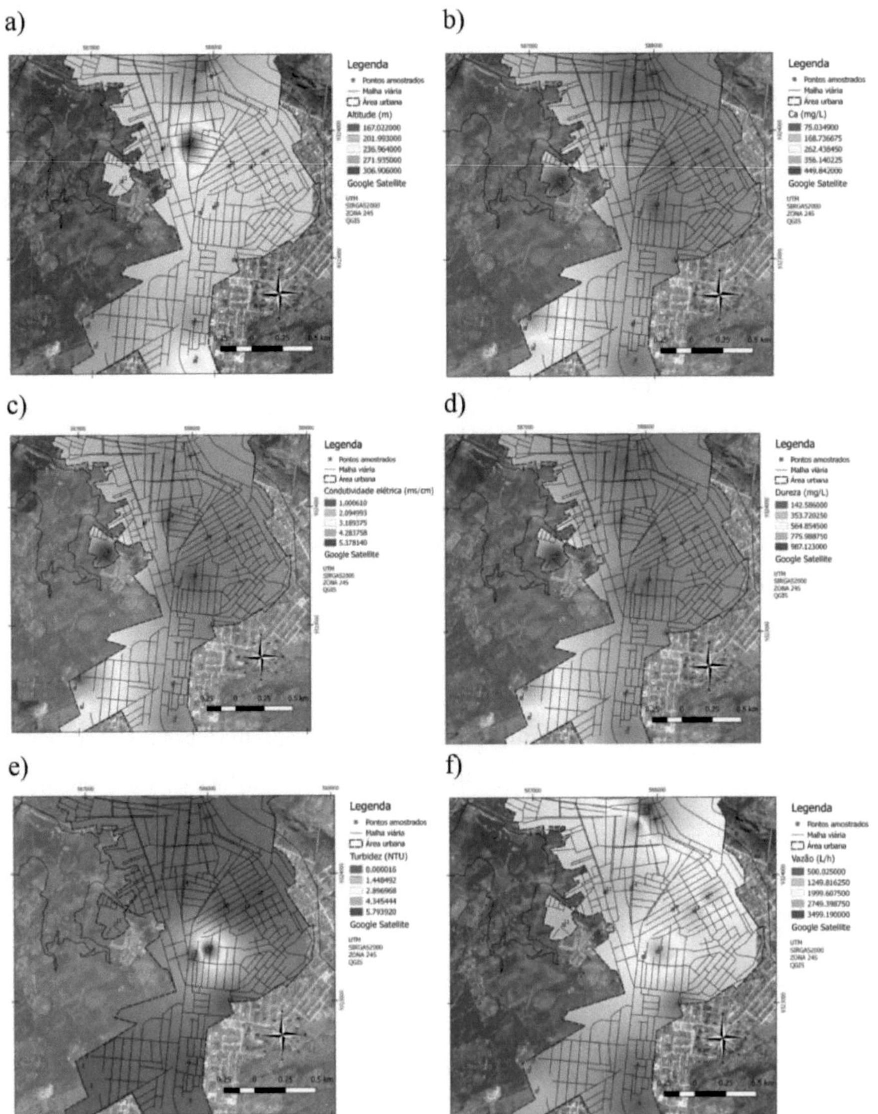

Figure 10 - Thematic isoline maps of the variables: a) Altitude; b) Calcium; c)

Electrical Conductivity; d) Hardness; e) Turbidity and f) Flow Source: Prepared by the authors

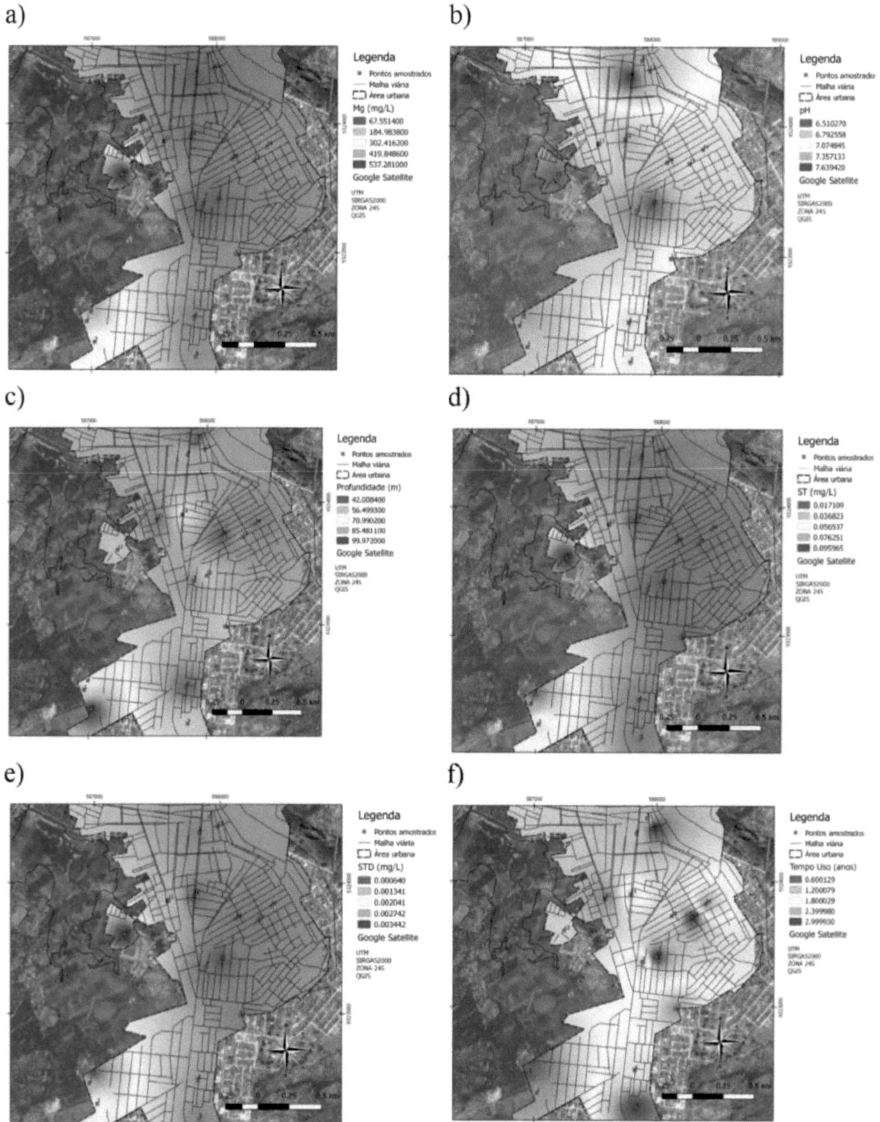

Figure 11 - Thematic isoline maps of the variables: a) Magnesium; b) Hydrogen potential; c) Depth; d) Total Solids; e) Total Dissolved Solids and f) Time of Use.

Source: Prepared by the authors

Figures 11a, 11d and 11e show the variables magnesium, total solids and total dissolved solids, where the point with the highest concentration of all three variables was well 12, with a value of 537mg/L for magnesium, 0.09mg/L for total solids and 0.003 for total dissolved solids, located in the Sao Judas Tadeu district.

The amount of magnesium is associated with the amount of calcium present in the water, both are together, with magnesium being more difficult to detect because it is present in smaller quantities than calcium, values that can be confirmed by Table 2.

Figures 11b, 11c and 1!f show the distributions of the pH, depth and time of use concentrations of the water, respectively, where the highest concentrations were found in well 6 for pH, well 3 for depth and wells 9, 10, 13 and 17 for time of use. The greatest depth found was 100 meters, showing the difficulty of obtaining water with the lack of rain. The time in use was between 2.4 and 3 years, which is when drilling began, given the start of the drought.

There was a higher percentage of shallow wells, dug by a mechanized drilling system, up to 55 meters deep, whose water was collected by pumping. Attention was drawn to the high number of households using a sewage disposal system via a pit/sump, which are often located at short distances from the collection well to the disposal pit. This potentially puts the quality of groundwater at risk through infiltration and percolation through fissures, or exposure in aquifer recharge areas.

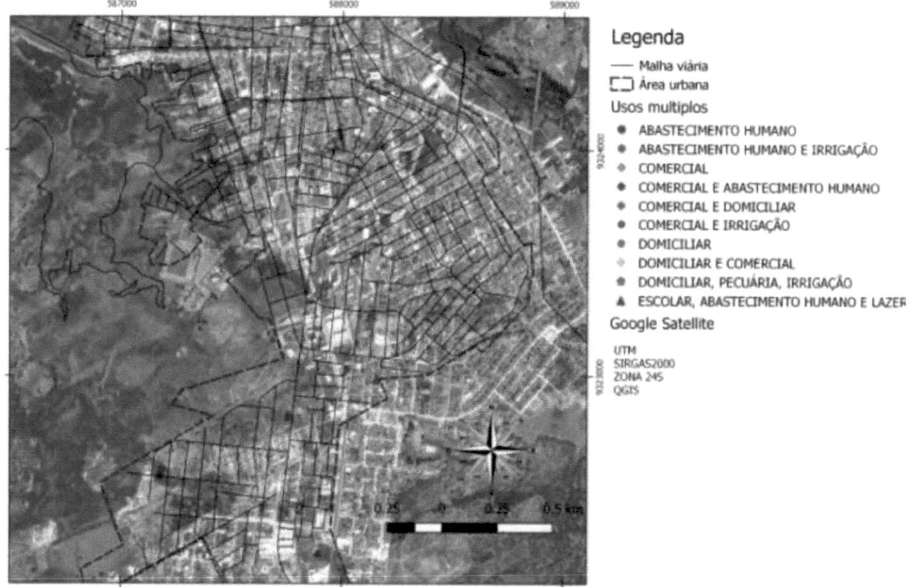

Figure 12 - Thematic map of the multiple uses of groundwater in the urban area of Pau dos Ferros-RN

Source: Prepared by the authors

5 FINAL CONSIDERATIONS

For the parameters evaluated, the groundwater showed extreme values for hardness, calcium and magnesium, as well as values within the drinking water standards for total solids, total dissolved solids and pH, indicating the presence of hard water, which restricts its use for some multiple uses.

The groundwater showed high levels of calcium, magnesium and hardness, and is not suitable for immediate consumption by the population or animals. It could be contaminated by an excess of some mineral (metallic ion), and it is advisable to have these waters analyzed by specialized bodies in order to provide quality water.

The water does not meet the potability standards recommended in ordinance 2.914/11. Therefore, human consumption of this water may represent a risk and aggravation to health.

The statistical data was of great importance in order to have a more concrete idea of the quality of the water in these wells, so that a measure could be devised to reduce the amount of excess ions, so that the population can enjoy quality water.

From the statistical data, it can be seen that the well that remained outside the drinking water standards according to the Ministry of Health's Ordinance 2914/2011 was well 12, located in the Sao Judas Tadeu neighborhood, where there were high amounts of calcium and magnesium, as well as hardness, turbidity and electrical conductivity.

The spatialization of the wells was satisfactory, managing to relate well the distribution and quality of the water in the wells, as well as their physiographic aspects.

The exploitation of groundwater through the drilling of artesian wells tends to continue increasing in a disorderly manner and without concern for the sustainability of this resource, since there is no forecast of the main source of recharge of this water.

Groundwater plays an important role and, in many cases, is vital for the supply of drinking water. For this reason, it is recommended that it be protected by eliminating the causes of possible contamination, and that it be used rationally to promote the sustainability of this resource, as well as reducing the risk of water contamination to a significant level.

REFERENCES

National Water Agency (ANA). *Overview of surface water quality in Brazil*, 2012 Available at:<

http://arquivos.ana.gov.br/institucional/sge/CEDOC/Catalogo/2012/PanoramaAguasSuperficiaisPortugues.pdf)>, accessed on: 18 Aug. 2016.

National Water Agency (ANA). *Groundwater,* 2009. Available at:

<http://www.uniagua.org.br/public_html/website/estudo_aguas_subterraneas.pdf>. Accessed on: October 20, 2016.

Brazilian Groundwater Association (ABAS). Hydrogeological Studies. Available at:

<http : //www.abas. org/estudo s_termo s .php>, accessed on: October 31, 2016.

ARAÙJO, R. Aquifers: inaccessible in 60% of RN.Tribuna do Norte, Rio Grande do Norte, Oct. 10, 2015. Available at:

<http:// http://www.tribunadonorte.com.br/noticia/aqua-feros-inacessa-veis-em-60-do- rn/326729>. Accessed on: August 7, 2016.

BELTRÂO, B. A.; ROCHA, D. E. G. A.; MASCARENHAS, J. C.; JUNIOR, L. C. S.; PIRES, S. T. M.; CARVALHO, V. G. D.; Projeto Cadastro de fontes de abastecimento por Âterrânea Estado do Rio Grande do Norte: Diagnòstico do Municipio de Pau dos Ferros. Pau dos Ferros: Editora Eletrônica, 2005. p.11.

BRAZIL. Ministry of Mines and Energy. Department of Mines and Metallurgy. CPRM - Geological Survey of Brazil [CD-ROM]. Geology, tectonics and mineral resources of Brazil. Geographic Information System - GIS. Maps on a scale of 1:2,500,000. Brasilia: CPRM, 2001. Available on 04 CDs.

BRAZIL. Ministry of Mines and Energy. Department of Mines and Metallurgy. CPRM - Serviço Geològico do Brasil.2004 Cadastro de Fontes de Abastecimento por Agua Subterrànea Project. Recife: CPRM (in press).

BRAZIL. Ministry of Health. Ordinance 2.914, of December 12, 2011. Provides for procedures to control and monitor the quality of water for human consumption and its standard of potability. Brasilia: Ministry of Health, 2011. Available at:

<http://bvsms.saude.gov.br/bvs/saudelegis/gm/2011/prt2914_12_12_2011.html> . Accessed on: August 10, 2016.

COSTA, A. M. B.; MELO, J. G.; SILVA, F. M. Aspectos da salinizaçao das águas do aquifero cristalino no estado do Rio Grande do Norte, Nordeste do Brasil. Aguas Subterrâneas, v.20, n.1, p.67-82, 2006.

COSTA, A. M. Zoneamento Hidroquimico do Aqüifero Cristalino do Rio Grande do Norte, PPGEO/CCET/UFRN, Master's dissertation, 2002.

CUSTÓDIO, E; LLAMAS, M.R. Hidrologia subterrànea. 2° ed. Barcelona: Ediciones Omega, 2v, 1983. Available at:

<http://www.bdtd.ufpe.br/tedeSimplificado//tde_busca/arquivo.php?codArquivo =1071> Accessed on: 07 Jan. 2010.

National School of Public Administration (ENAP). Socioeconomic Evaluation of Projects Program, 2015. Available at: <

http://antigo.enap.gov.br/index.php?option=com_content&task=view&id=2040 &Itemid=2042>. Accessed on: 20 Nov. 2016.

EMPARN. Precipitation data for RN. Available at:

<http://www.emparn.rn.gov.br/contentproducao/aplicacaoemparn/arquivos/metereologa/precipitacao.asp> Accessed on: 08 Nov. 2016.

EMBRAPA. Brazilian Agricultural Research Corporation, 2008. Regionalization of groundwater quality parameters for irrigation in the state of Sergipe. Available at:

<http://www.cpatc.embrapa.br/publicacoes_2008/bp_36.pdf> Accessed on: 20 Nov. 2016.

FERNANDES, A. Fitogeografia brasileira: fundamentos fitogeogràficos. Cearà: Ediçôes UFC, 2007.

FEITOSA, F. A. C.; VERiSSIMO, L. S.; AGUIAR, R. B.; SANTOS COLARES, J. Q.; MORAES, F.; GALVÂO, M. J. T. G.; COSTA FILHO, W. D.; OLIVEIRA, L. T.; VIEGAS, J. C.; Hydrogeological studies of sedimentary basins in the semi-arid region of northeastern Brazil. Sao Paulo: Editora Eletrônica, 2004. 84 p. Available at:

<https://aguassubterraneas.abas.org/asubterraneas/article/view/22996>. Accessed on: September 18, 2016.

FOSTER, S. 1993. Determining the risk of groundwater contamination: a method based on existing data. Geological Institute, Sao Paulo.

GOMES, F. P. Modern statistics in agricultural research. Piracicaba, PATAFÓS, 1984, 160p.

HIRATA, R. Water resources. In: Deciphering the earth. Wilson Texeira et al. (org.) 2. Reprint, Sao Paulo: Oficinas de textos, 2003.568p

Brazilian Institute of Geography and Statistics (IBGE). Cities - Data for the municipality of Pau dos Ferros, 2010. Available at:

<http://cidades.ibge.gov.br/xtras/perfil.php?lang=&codmun=240940&search=ri o-grande-do-norte|pau-dos-ferros|infograficos:-informacoes-completas>. Accessed on: October 25, 2016.

Brazilian Institute of Geography and Statistics (IBGE). Cities - Data for the municipality of Pau dos Ferros, 2014. Available at:

< http://cidades.ibge.gov.br/xtras/temas.php?lang=&codmun=240940&idtema=15 5&search=rio-grande-do-norte|pau-dos-ferros|estatisticas-do-cadastro-central- de-empresas-2014 >. Accessed on: October 25, 2016.

Brazilian Institute of Geography and Statistics (IBGE). Cities - Data for the municipality of Pau dos Ferros, 2015. Available at: < http://cidades.ibge.gov.br/xtras/temas.php?lang=&codmun=240940&idtema=15 6&search=rio-grande-do-norte|pau-dos-ferros|ensino-matriculas-docentes-e- rede-escolar-2015 >. Accessed on: October 25, 2016.

Brazilian Institute of Geography and Statistics (IBGE). Foundation of the Brazilian Institute of Geography and Statistics. Rio de Janeiro: IBGE. 2000.

Brazilian Institute of Geography and Statistics (IBGE). National Survey of Basic Sanitation, 2000. Available at:

<http://www.ibge.gov.br/home/estatistica/populacao/condicaodevida/pnsb/p nsb.pdf>. Accessed on: August 7, 2016.

Institute for Sustainable Development and the Environment of Rio Grande do Norte - IDEMA. Profile of your municipality. Natal-RN. 2009. Available at: WWW.idema.rn.gov.br. Accessed on: October 25, 2016.

Rio Grande do Norte Water Management Institute (IGARN). Agua Azul Program, 2016. Available at: <

http://www.igarn.rn.gov.br/Conteudo.asp?TRAN=ITEM&TARG=24385&ACT =&PAGE=0&PARM=&LBL=Programas>. Accessed on: 08 Nov. 2016.

KEMPER, K. E. 2007. Instruments and Institutions for Groundwater Management. In: Giordano, M. and Villholth, K.G. The Agricultural Groundwater Revolution: Opportunities and Threats to Development. CABI Publishing Series. Washington (DC), 419p.

LARINI, M. M. Evaluation of the use of groundwater in the metropolitan region of Londrina-PR and comparison of the use of surface water sources for public supply. Monograph, UTFPR, Londrina, PR, 2013.

MAFFEI, F.; CARBONE, F.; CANTELLI FORTI, G.; BUSCHINI, A.; POLI, P.; ROSSI, C.; MARABINI, L.; RADICE, S.; CHIESARA, E.; HRELIA, P. (2009). Drinking water quality: an *in vitro* approach for the assessment of cytotoxic and genotoxic load in water sampled along distribution system. Environment International 35, pp. 1053-1061.

Ministry of the Environment (MMA). Groundwater: A resource to be known and protected, 2007. Available at:

< http://www.agrolink.com.br/downloads/%C3%81GUAS%20SUBTERR%C3%82NEAS.pdf >. Accessed on: 08 Nov. 2016.

OLIVEIRA, L. I.; LOUREIRO, C. O. 1998, Contamination of aquifers by organic fuels in Belo Horizonte: Preliminary evaluation. In: X Congresso Brasileiro de Aguas Subterrâneas, 2000. Available at:

< https://aguassubterraneas.abas.org/asubterraneas/article/view/22287/14630>. Accessed on: October 15, 2016.

SANTOS, C. M. A. Descriptive Statistics - Self-learning Manual. Lisbon: Ediçôes Silabo, 2007.

SILVA, R. C. A.; ARAUJO, T. M. Quality of groundwater in urban areas of Feira de Santana (BA). Editora Eletrônica, Feira de Santana, n. 1, p.1019-1028, 08 sep. 2003. Available at:

<http://www.scielo.br/pdf/csc/v8n4/a23v8n4>. Accessed on: 09 Oct. 2016

SILVA, R. C. A. 1999. Giving up the right to consume treated water: Feira de Santana - BA. Monograph presented to the specialization course in Health Law. Department of Health, State University of Feira de Santana.

SILVA, M. A. R. Economics of natural resources. In: MAY, P.; LUSTOSA, M. C.; VINHA, V. (Orgs.). Environmental economics: theory and practice. Rio de Janeiro: Editora Campus, 2003.

Secretary of the Environment and Water Resources of Rio Grande do Norte (SEMARH). Profile of your municipality. 10. ed. Natal: Mossoró, 2008. 24 p.

TEIXEIRA, W. Deciphering the Earth. Sao Paulo: Companhia Editora Nacional, 2008. 557 p.

TUNDISI, J. G. Recursos Hidricos. Multiciência, Sao Carlos - SP, v. 1, n. 1, p.10-11, 2003.

TUINHOF, A. Sustainable Groundwater Management: Concepts and Tools. Groundwater Monitoring: Requirements for managing aquifer response and quality threats. GWMate. Briefing Note Series Briefing Note 9. World Bank. Global Water Partnership Associate Program. 10p. 2004.

UNEP/WHO. Water Quality Monitoring - A Practical Guide to the Design and Implementation of Freshwater Quality Studies and Monitoring Programs. World Health Organization. Geneva. 1996.

Printed by Books on Demand GmbH, Norderstedt / Germany